Advanced Hydroponics Technologies

For Growing Fruits and Vegetables

Roby Jose Ciju

All Rights Reserved. No parts of this publication may be reproduced, stored in a retrieval system, or transmitted, in any form or by any means, electronic, mechanical, photocopying, recording, or otherwise, without the prior permission of agrihortico

© 2012 **AGRIHORTICO**

CONTENTS

1. Introduction 3
2. Growing Media for Hydroponics 4
   2.1 Perlite 4
   2.2 Rockwool 4
   2.3 Expanded Clay Pebbles 4
3. Types of Hydroponics Systems 5
   3.1 Aggregate Systems 5
      3.1.1 Ebb and Flow Hydroponics 5
   3.2 Deep Water Culture (DWC) Systems 6
      3.2.1 NFT Hydroponics 6
      3.2.2 BubblePonics 6
   3.3 Aeroponics 7
   3.4 Drip System 7
4. Advanced Hydroponics Grow Systems 8
   4.1 Overview 9
   4.2 Propagation Systems 10
      4.2.1 Starter Tray 10
      4.2.2 Humidity Dome 10
      4.2.3 Seedling Heat Mat 10
      4.2.4 Heat Mat Thermostat 10
   4.3 Growing Systems 11
      4.3.1 Temperature and Humidity Regulation 11
      4.3.2 Ventilation or Air Flow System 12
      4.3.3 Oxygenation or Aeration System 12
      4.3.4 Air Filtration System 12
      4.3.5 Lighting System 12
      4.3.6 CO_2 Application System 14
      4.3.7 Nutrient Management System 14
      4.3.8 Ozone Generating System 18
      4.3.9 Automated Time Management System 18
      4.3.10 Water Management System 18
      4.3.11 Additional Plant Support System 19

	4.3.12	Fogging System	19
	4.3.13	Pest and Disease Management	19
5		Product Study	21
	5.1	Technology – Simple Ebb n Flow Hydroponics	21
	5.2	Technology – NFT Hydroponics for Hobby Gardening	21
	5.3	Technology: NFT Hydroponics for Growing Herbs	22
	5.4	Technology: Fusion of BubblePonics and Ebb n Flow	22
	5.5	Technology: Aeroponics	23
	5.6	Technology: Automated All-in-One Triple Chamber Grow Cabinet	23
	5.7	Technology: Automated All-in-One Dual Chamber Grow Cabinet	25
	5.8	Technology: Hydroponics Grow Tent	25
	5.9	Technology: Vertical Hydroponics (VerticalPonics)	26
	5.10	Technology: Fully Automated All-in-One Hydroponics Trailers	28
	5.11	Technology: Hydroponics Greenhouses	29
6		Annexure	31
		Annexure 1: Market Price of Some Ready-to-Install (fully automated) Hydroponics Systems	31
		Annexure 2: Market Price of Various Components of a Hi-Tech Hydroponics System	32
7		References	33

1. Introduction

Today the term *'hydroponics'* has become synonymous with *'soilless production of crop plants'* though the term itself means that it is water (*'hydro'* means *'water'*) that is at work (*'ponics'* means *'labor'* or *'work'*). The term 'hydroponics' may seem to represent a sophisticated process but in reality, *hydroponics is a simple process of crop production*. The only difference of a hydroponics crop production from that of a traditional method is that **hydroponics makes use of a nutrient solution as plant growing medium instead of soil**.

Hydroponics under controlled environmental conditions (i.e. *greenhouse hydroponics*) is more successful than that of outdoor hydroponics. When higher yields per unit area and higher productivity per plant are desired, greenhouse hydroponics is preferred than outdoor hydroponics.

Hydroponics is based on the principle that plant growth in a traditional soil-based production system is not dependent on the soil rather it is dependent on the nutrients and moisture present in the soil. So, if the plant nutrients and moisture required for the plant growth are provided through any other medium other than the soil, plants can still have a natural growth. Therefore in a hydroponics system, ideal nutrient and moisture requirements of the plants are fulfilled through a **water culture** or **solution culture** under ideal environmental conditions. In other words, ***hydroponics is soilless crop production system under controlled environmental conditions.***

Not all plants are suitable for hydroponics systems. Generally, seasonal annual plants can successfully be grown using hydroponics. Both *quality* and *quantity* of the produce are increased in hydroponics growing systems.

2. Growing Media for Hydroponics

Different types of growing media like rockwool, vermiculite, perlite, etc may be used in a hydroponics system. However, most popular and highly recommended growing media are perlite, rockwool and expanded clay pebbles. This is because these media maintain a neutral pH always and *do not alter the pH of nutrient solution* used in a hydroponics system. Since pH of the nutrient solution has a strong influence on plant growth, a growing medium should be selected based on its pH reaction.

2.1 Perlite

Perlite, a porous growing medium derived from volcanic rocks, is available in the market as *grow bags*, thin plastics sleeves filled with perlite stones or as loose, white-coloured porous stones. Individual grow bags are used as such to grow individual plants.

2.2 Rockwool

Rockwool, a fibrous growing medium derived from basalt rocks, is available in the market as *rockwool cubes* or as *rockwool slabs*. Rockwool cubes are mainly used for propagation purposes while rockwool slabs are used for transplanting and vegetative growing purposes.

2.3 Expanded Clay Pebbles

Expanded clay pebbles are available in the market as loose clay stones. Clay pebbles are highly recommended in 'ebb and flow' hydroponics systems because of their high degree of capillary action and neutral pH.

3. Types of Hydroponics Systems

Major types of hydroponics systems are Aggregate Systems, Deep Water Culture (DWC) Systems, Aeroponics and Drip System of Hydroponics.

3.1 Aggregate Systems

Aggregate system of hydroponics makes use of aggregates or solid *growing media to provide support to the root systems* of growing plants. Growing media generally used for aggregate systems are rockwool, perlite or expanded clay pebbles.

In a simple aggregate system, a tray filled with a growing medium is used for planting seedlings and/or clones. This plant tray is connected to a reservoir tank filled with nutrient solution. A water pump is attached to the reservoir tank to inject the nutrient solution from the tank to the plant tray. Plant tray and reservoir tank are connected in such a way that nutrient solution is directly pumped to the root zone of the growing seedlings. Nutrient solution is pumped until the aggregate growing medium that holds the plant roots is saturated with the nutrient solution. At this saturation point, the growing medium is no more flooded with the nutrient solution but left undisturbed for a few hours until the excess solution is drained out. Proper drainage is necessary for the roots to have proper aeration. After the draining process, the growing medium is again flooded with the nutrient solution. This *flood* and *drain* process is repeated at regular intervals. That is why aggregate hydroponics is generally known as '*flood and drain*' hydroponics.

3.1.1 Ebb and Flow Hydroponics

A popular aggregate hydroponics system is Ebb and Flow (*Flood and Drain*) system. Most of the Ebb and Flow systems are **recirculation hydroponics** systems as the nutrient solution drained from the growing media is recycled or recirculated through the growing system again and again. One of the disadvantages of recirculation hydroponics is that pH levels of the nutrient solution tend to be unstable due to the recirculation process and hence constant pH monitoring is required.

3.2 Deep Water Culture (DWC) Systems

DWC systems use only water culture or nutrient solution to grow the plants. These systems *do not use any aggregate growing media as plant root support system*. The plant roots are totally suspended in the nutrient solution. A tray made of plastic or Styrofoam boards or similar materials that float on the surface of the solution is used to support the plant above the solution. Holes are provided on the tray so that the roots are inserted into the solution while the shoots stand on the tray growing upwards. In DWC systems, the nutrient solution needs to be aerated or bubbled continuously by using an air pump and air stones. Nutrient solution should be changed regularly and kept at constant level in the reservoir tank. Best examples of popular DWC systems are *Nutrient Film Technology* (NFT) and *BubblePonics*.

3.2.1 NFT Hydroponics

NFT is one of the most popular water culture hydroponics systems used for growing herbaceous plants and leafy vegetables such as lettuce and spinach.

3.2.2 BubblePonics

BubblePonics or *bubble hydroponics* is water culture hydroponics where *constant oxygenation* or aeration of the plant root zones is required for the healthy growth of the plants.

3.3 Aeroponics

Aeroponics is a DWC system where plant roots are suspended in a *nutrient mist* supplied directly to the plant roots by using vaporizers.

3.4 Drip System

Drip system of hydroponics is similar to *ebb and flow system* in its operative principle with the major difference being here the nutrient solution is not flooded to the plant root zone instead applied carefully to the plant root zones by using *a controlled dripping mechanism*.

4. Advanced Hydroponics Grow Systems

Modern advanced hydroponics grow systems *combine one or more hydroponics systems into a highly efficient grow system unit* to ensure higher yields and high quality of the produce. This fusion of different hydroponics technologies into *a fully automated hi-tech hydroponics grow system* has made it possible to produce hydroponics crops on a commercial scale.

Fully automated advanced hydroponics systems are now available from market leaders such as SuperCloset, American Hydroponics, and Universal Hydro. Many different types of hydroponics systems are available in the market for different growing purposes. Some of the advanced hydroponics systems that are worth mentioning here are,

1. Fully automated hydroponics trailer greenhouses
2. Hydroponics greenhouse packages with hydroponics systems
3. Fully automated grow cabinets and grow boxes
4. High quality grow tents
5. Fully automated VerticalPonics systems
6. Fully automated aeroponics systems

Hydroponics growers may purchase either a fully automated unit at a premium price or they may go for individual system components to be assembled into a fully automated functional hydroponics unit.

4.1 Overview

Since plant growth is based on **FIVE** fundamental growth requirements such as *Light*, *Air*, *Water*, *Nutrition*, and *Climate*, advanced hydroponics systems come with fully automated systems for light, air, water, nutrition and climate management. Major components of a hi-tech hydroponics grow system includes

1. Light management system

2. Air management system : air filtration system, CO_2 application system, ventilation or internal air flow system and ozone generating system

3. Water management system - water filtration system and water pumping system

4. Nutrient management system – nutrient application system and nutrient monitoring system

5. Climate control system – for humidity and temperature control

6. Miscellaneous - automated time management system, plant support system etc

A standard hydroponics grow system is divided into **TWO** areas: *a propagation area* (propagation room) and a *grow area* (grow room). Propagation area is where propagation takes place while grow area is exclusively dedicated for vegetative growing purposes.

A standard hydroponics system includes grow trays, reservoir tanks, plant pots, plant nutrient kits, pH test kit, digital TDS meter, water pumps, air pumps and air stones, air diffusers, growing media kits, starter materials (seeds, clones), and a plant starter tray. The system should have activated carbon filters, duct mufflers or silencers, and a light proof system (light baffles and light leak seal). Safety of operations is ensured in a hi-tech hydroponics system by providing provisions for fire protection, and insulation.

4.2 Propagation Systems

Propagation room is used for propagation purposes. Propagation may include seed propagation and/or vegetative propagation (i.e. propagation by means of cuttings, suckers, rhizomes, tubers or any other vegetative part of the plant that is used for propagation). Propagation room is specifically designed for facilitating *propagation process* by providing optimum temperature, light and humidity conditions. A *standard propagation system* available with the hi-tech hydroponics system includes a starter tray, a humidity dome, a seedling heat mat, a heat mat thermostat and a growing media kit.

4.2.1 Starter Tray

A starter tray is used for seed germination and/or cloning (vegetative propagation). Seedlings and/or clones are grown here until they attain transplanting age. Starter trays are made up of durable plastic and come with a reservoir tank in which nutrient solution is prepared to be supplied to the saplings and/or clones.

4.2.2 Humidity Dome

A humidity dome is used for providing optimum humidity conditions for starting planting materials.

4.2.3 Seedling Heat Mat

Heating the root zone of germinating seeds and/or growing clones is important for achieving their optimum growth. Seedling heat mat is used for this purpose in a hydroponics system.

4.2.4 Heat Mat Thermostat

Heat mat thermostat is used to regulate the temperature of the heat mats.

> **Tips for Successful Propagation Process**
>
> *It is advised to start planting materials in the hydroponics system itself. This helps avoid transplant stress for the plants. Propagation within the hydroponics system will also help the growers obtain disease-free and pest-free starting materials*

Seed propagation on a starter tray: Seeds are sown directly on the starter tray that is filled with a growing medium. After sowing, tray is covered with a lid. The growing medium is constantly kept moist by pumping the nutrient solution kept in a reservoir tank. Rockwool is the most popular growing media used in the hydroponics. Rockwool cubes with a hole in the center are popularly used for seed propagation. These cubes are soaked in water or nutrient solution before placing seeds into the hole. Cubes are kept moist until the germination process is completed and seedlings are ready for transplanting.

4.3 Growing Systems

Seedlings and/or clones, when they are ready for transplanting, are taken to the grow room. A grow room has a perfectly balanced environment with proper ventilation, humidity, temperature and light management system.

4.3.1 Temperature and Humidity Regulation

A controlled environment is recommended for advanced hydroponics because such a system provides shelter, and stress-free environment for the plants. Temperature of the growing environment may be monitored regularly by using a thermometer. A hygrometer may be used to measure the humidity inside the growing environment.

4.3.2 Ventilation or Air Flow System

A standard air flow system includes 6" internal circulation fan with an air flow of 400 cubic feet per minute (cfm), and a thermometer and a hygrometer. A well-designed hydroponics system has cross air flow system to ensure adequate aeration around the plants and their root zones.

4.3.3 Oxygenation or Aeration System

In hydroponics, the roots of the growing plants should be aerated at regular intervals because plant roots need oxygen in order to survive. This oxygenation of roots may be carried out by using highly efficient air pumps and air stones.

> *Tips for Healthy Plant Growth*
>
> *In hydroponics, care should be taken not to expose the roots of the growing plants to the light. Root exposure to light may induce growth of algae and thus contaminate the growing medium.*

4.3.4 Air Filtration System

Industrial grade carbon-based air filtration system is very effective in hydroponics for air purification. Carbon filters effectively eliminates undesirable odors from the grow system.

4.3.5 Lighting System

A standard lighting system includes 600 watt digital ballast, HPS lights, MH lights, light tube fixture, reflector and a timer.

4.3.5.1 High-Intensity Discharge (HID) Metal Halide Light System

It is the safest and economical way of providing light for plants. Two popular HID lighting systems are Metal Halide (MH) and High-Pressure Sodium (HPS). HPS systems should be used with MH systems to achieve a balanced light spectrum within a hydroponics system.

Difference between MH and HPS light systems is given below:

MH Light System	HPS Light System
MH produces blue-white light spectrum that is excellent for vegetative growth of the plants.	HPS produces orange- red light spectrum that is excellent for fruiting or flowering of the plants.

4.3.5.2 LED Grow Light

LED lights are highly energy efficient and economical. LEDs are low temperature way to increase the amount of light that plants receive. A LED light can replace the standard 600 watt HID grow light and thus reduces energy consumption. LEDs are most suitable for fully automated grow boxes and grow cabinets.

4.3.5.3 Reflectors

Light distribution and coverage within the system can be adjusted by installing panels and reflectors.

Light Placement

Placement of the lights should be directly related to the intensity of the light required by the plant. If more light intensity is required, place the light close to the plants but not too close to burn the leaves. Adjustable lighting system may be used to adjust the light according to the plant requirement.

> *Significance of Light Energy and CO_2 in Hydroponics*
>
> *Plants need light energy for various purposes, major being photosynthesis and transpiration. During photosynthesis, plants produce carbohydrates (foods) using light energy, carbon dioxide and water. In an enclosed hydroponics system, artificial lighting system and CO_2 application system may be used to provide the light and CO_2 needed by the plants*

4.3.6 CO$_2$ Application System

CO$_2$ forms an integral part of a plant growth system and therefore it is important that CO$_2$ should be applied in a hydroponics system for healthy plant growth. Generally CO$_2$ is administered to the plants through a tank application process. Alternatively, a CO$_2$ kit may be used to ensure automated CO$_2$ delivery. This kit contains a solenoid, regulator, timer and injection tube which is connected to a CO$_2$–filled bucket to ensure automated CO$_2$ delivery. A refill bucket is also available in the market so that an empty CO$_2$ bucket can be refilled on time.

4.3.7 Nutrient Management System

For healthy plant growth, a plant needs both macronutrients and micronutrients (trace elements). Major macronutrients include Nitrogen (N), Phosphorous (P), Potassium (K), Calcium (Ca), Magnesium (Mg), and Sulfur(S). Trace elements are Iron (Fe), Manganese (Mn), Boron (Bo), Zinc (Zn), Copper (Cu), and Molybdenum (Mb). Nutrient formulas containing all these nutrients in correct proportions are available in the market as *hydroponic nutrient mixes*.

Tips for Successful Nutrient Application

- ✓ *Whenever a nutrient solution is prepared, use a measuring cup to take correct quantities of nutrients*
- ✓ *Make sure that nutrients are taken in correct proportions*
- ✓ *If nutrients are in powder form, use warm water to dissolve them and mix well by stirring vigorously to get a homogenous solution*
- ✓ *Use a PPM (parts per million) measuring device (e.g.: nutrient monitor) to measure the concentration level of the nutrient solution*

While selecting a hydroponic nutrient mix for the plants, maximum care should be taken to choose the right product. This is because each type of plant has its own nutrient requirement. While applying nutrients, use each nutrient in right proportions and in accurate quantities. Generally, a measuring cup will also be available with the nutrient package so that nutrient may be applied in measured quantities.

Functions of Plant Nutrients and Corresponding Deficiency Symptoms

Nutrient	Function	Deficiency Symptoms
Nitrogen (N)	Protein synthesis and chlorophyll synthesis	Chlorosis (chlorophyll deficiency in plants) and stunted plants. Excess nitrogen leads to excessive vegetative growth and therefore delayed fruiting and ripening processes
Sulfur (S)	Protein synthesis	Small leaves, pale green leaves, and suppressed fruit formation
Phosphorous (P)	Synthesis of proteins, phospholipids, sugar phosphates, and nucleic acids	Premature leaf fall, poor fruiting, reduced copper and zinc availability
Potassium (K)	K helps formation of carbohydrates and proteins; and helps in transpiration regulation and photosynthesis	Chlorosis, stunted plants with numerous tillers, and little or no flowering
Calcium (Ca)	An important constituent for the formation of cell wall	Stunted plant growth and brown spots along the leaf margins
Magnesium (Mg)	Chlorophyll synthesis	Chlorosis
Iron (Fe)	Fe plays an active role in respiratory process, and chlorophyll synthesis	Chlorosis in a mottled pattern
Boron (Bo)	Bo role plays an active in carbohydrate breakdown	Stunted root growth; stem elongation, sterile flowers or lack of flowering, deformed fruits, and die-back of stems
Zinc (Zn)	Zn plays an active role in carbohydrate metabolism, CO_2 utilization, and phosphorus metabolism	Intervenal Chlorosis and reduced availability of Fe
Manganese (Mn)	Mn plays an active role as an activator of several enzymes of aerobic respiration	Chlorosis and reduced availability of Fe
Copper (Cu)	Cu is a chief constituent of ascorbic acid oxidase system and helps in achieving Carbon/Nitrogen balance in plants	Reduced vegetative and reproductive growth
Molybdenum (Mb)	Mb plays an active role in nitrogen fixation and nitrate reduction	Necrosis of leaf tissues

Preparation of Nutrient Solution

While preparing an ideal nutrient solution, pH, electrical conductivity (EC), temperature and total dissolved solids (TDS) of the solution should be measured as each of these parameters has an impact on the degree of nutrient absorption by the plant roots.

4.3.7.1 pH of Nutrient Solution

pH of the nutrient solution can have a great impact on the plant growth. Since every plant has a preferred pH range at which plant nutrients become available to its growth, solutions having too low or too high pH should be avoided in a hydroponics system. pH of nutrient solution should be checked regularly by using any of the pH devices available in the market. Cheap pH devices such as a *pH control kit* and *pH pen* may serve this purpose for those who are looking for cost effectiveness. A *pH meter* may be a costly device as compared to a pH control kit but provides instant reading.

4.3.7.2 Electrical Conductivity (EC) of Nutrient Solution

EC refers to '*electrical conductivity*' or flow of electric current through the nutrient solution. EC and concentration of the nutrient solution is proportionately correlated. i.e. when the concentration of nutrients is higher in the solution, EC will be higher and vice versa.

EC Meter

EC meter is used to measure the electrical conductivity of the nutrient solution. EC meter records the reading in either *micromhs per centimeter* (uMho/cm) or *microsiemens per centimeter* (uS/cm). The temperature of the nutrient solution affects the reading of the EC meter. Hence it is recommended that ***EC should be measured at 25^0 C always***. If the temperature of the nutrient solution is above 25^0 C, the EC reading will be higher, even though concentration of the solution remains same. If the temperature of the nutrient solution is below 25^0 C, EC reading will be on the lower side.

4.3.7.3 Total Dissolved Solids (TDS) of Nutrient Solution

TDS refers to the *total dissolved solids* present in the nutrient solution.

TDS Meter

A TDS meter is used to measure TDS level of the nutrient solution. The meter reading is shown in *milligrams per liter* (mg/l) or *parts per million* (ppm).

Nutrient Monitoring Process

Nutrient monitoring process is very important in a hydroponics growing system. For hi-tech hydroponics growing systems, nutrient monitoring and recording of the results is recommended on a daily basis. Regular testing of the nutrient solution for pH, EC, and TDS helps growers ensure that **plants are being fed with right nutrients at right concentration**. It also helps monitor the salt levels of the nutrient solution at every phase of crop production. Thus growers can take appropriate corrective measures if the salt levels rise unexpectedly resulting in a '*salt build-up*' in the growing system. A **24x7 *Nutrient Monitoring Device*** is available in the market for nutrient monitoring purpose. With such advanced technologies, constant nutrient monitoring in a hydroponics system is now possible throughout the day.

> ***Corrective Measures Recommended for Salt Build-Up***
>
> *Two important corrective measures recommended for eliminating problems associated with salt build-up is regular flushing of the growing medium with a fresh nutrient solution or replacing the nutrient solution with a fresh one.*

Advanced Nutrient Testing

In addition to the regular nutrient monitoring process, advanced leaf nutrient analysis is also recommended for commercial hydroponics systems. Leaf tissue analysis is the only way to determine exact nutrient composition of the plants.

Nutrient Test Kit

Nutrient test kits are available in the market to test the essential macronutrients (NPK) present in the leaf tissues.

4.3.8 Ozone Generating System

An ozone production system is also recommended in a hi-tech hydroponics system. Ozone system generates ozone which destroys all harmful microorganisms present in the growing environment.

4.3.9 Automated Time Management System

Timers are used for automated turning on and off hydroponic pumps, lighting systems, ventilation fans, CO_2 controller, and ozone generators.

4.3.10 Water Management System

Water Pumps: Powerful water pumps are used for water supply in hydroponics systems. These pumps are designated with 24-hour timers for regulating water pumping at periodic intervals.

Water Filters: In a hydroponic system, purity of water is very important. Generally RO (Reverse Osmosis) water filter is recommended to ensure water quality.

4.3.11 Additional Plant Support System

Additional support system such as net trellis, plant stakes, strings, and clips should also be provided for the plants.

4.3.12 Fogging System

Foggers or fogging equipments may be used to produce tiny fog particles for treating the hydroponics greenhouses or hydroponics grow systems. Good quality foggers produce high quality fog and ensure uniform fog distribution throughout the hydroponics greenhouse or grow system.

4.3.13 Pest and Disease Management

Since hydroponics systems are under constant monitoring and a controlled environment is used in most cases, pests and diseases are not a serious problem. Most of the pests and diseases are naturally eliminated in a clean and sanitary growing environment.

> *Tips for Successful Disease-Pest Management*
> - ✓ *Plant only healthy plants; remove all unhealthy plants from the system*
> - ✓ *Remove all dead and decaying leaves and plant matter*
> - ✓ *If there are any incidences of pests and diseases, use only biological control measures as biological controls are highly effective in closed growing systems*

If pest and disease problems are found, then organic (eco-friendly) control measures are advised. Some of the organic pest-disease management practices that may be adopted by a hydroponics grower are,

1. Use of bacterial cultures such as that of *Bacillus subtilis* for disease-free environment
2. Use of beneficial insects (natural enemies) to control harmful insects of the crops

3. Use of neem oil as an insecticide: neem oil affects chewing and sucking insects and does not harm beneficial insects

4. Use of neem oil as an effective fungicide

Grow Goggles

Specifically designed eye glasses are recommended for growers while working in a hydroponics system. Since a hydroponics system is lighted by artificial lighting system, using eye glasses reduce sodium glare and protect eyes from Ultra Violet and Infra Red rays emitted by HPS lighting.

Reserve Tank with Float Valve

Some grow systems come with an additional reserve tank which can link to a float valve inside the main reservoir tank to automatically replenish the nutrient solution as plants are kept on feeding with the solution.

> *Tips for Transplanting Process*
>
> *If seedlings are raised on rockwool cubes, then it is very easy to carry them to the grow chamber. Just raise the cubes along with seedlings and transplant them on the site meant for the purpose. Rockwool slabs and/or perlite grow bags are generally used for planting seedlings and/or cuttings in aggregate systems of hydroponics. In DWC systems, roots of growing seedlings are suspended in nutrient solution while shoots are supported above the solution by using a tray with holes. In aeroponics, roots are suspended in nutrient mist.*

5 Product Study

5.1 Technology – Simple Ebb n Flow Hydroponics

Brand: American Hydroponics, USA

Product: One Tray Econo System with Reservoir

Product Overview: One Tray Econo System comes with a plant tray, a reservoir tank and fitting accessories. Plants are grown in the tray while the reservoir tank is filled with the nutrient solution. Plants growing in the tray are supplied with the nutrient solution from the reservoir tank.

Price: $500.00

5.2 Technology – NFT Hydroponics for Hobby Gardening

Brand: North American Hydroponics, USA

Product: Hobbyist 33 system

Product Overview: The Hobbyist 33 System is ideal for growing 33 plants and comes with 33 Leg/Stand assemblies, 33 Grow Tube assemblies, 5 Liters Grow Rock, 33 Starter Cubes, and 33 Plant sites including: 33 Plant baskets, 33 Feed Lines, 33 Feed Stakes and 66 Pot Locks. The system also includes Convenience Valve Package, External Controls, Built-in Reservoir, Reservoir Lid, Hose Port and Pre-installed Submersible Pump.

Price: - $850.00

5.3 Technology: NFT Hydroponics for Growing Herbs

Brand: American Hydroponics, USA

Product: 2012 NFT (Nutrient Film Technology) Herb and Lettuce System

Product Overview: NFT channel pipes are made up of high quality HDPE (high density polyethylene). Channel pipes are provided with well-prepared plant sites where plants are grown while their roots are suspended into the pipe through which nutrient solution is pumped constantly.

Price: $8,695.00

5.4 Technology: Fusion of BubblePonics and Ebb n Flow

Brand: SuperCloset, USA

Product: Automated Bubble and Ebb n Flow Hydroponics Systems /Super BubbleFlow Buckets

Product Overview: Super BubbleFlow Buckets combines the technology of bubble hydroponics and ebb and flow hydroponics into an automated, recirculation hydroponic bucket system. The system provides consistent pH reading, TDS reading and uniform nutrient distribution throughout the entire growing system. The hyper-oxygenated environment in each individual bucket available for the roots of individual plants helps them grow vigorously.

Price: $595.00

5.5 Technology: Aeroponics

Brand: General Hydroponics, USA

Product: Automated Aeroponics Systems/AF60: 60 Site Aeroponics Systems

Dimensions: 5'1"L x 6'10"W x 2"H

Product Overview: This Automated Aeroponics Systems includes 40 gallon nutrient reservoir, six 6 feet grow chambers, water pump, injection manifold, support structures, 3 inch grow cups, growing media kit and a nutrient kit.

Price: $1,173.33

5.6 Technology: Automated All-in-One Triple Chamber Grow Cabinet

Brand: SuperCloset, USA

Product: Trinity 3.0 Grow Box

Dimensions of the Product – 72"w x 24"d x 72"h

Product Overview: Trinity 3.0 is a triple chamber unit with one cloning chamber for propagation purposes and two grow chambers for vegetative growth and flowering phases of plants. Cloning Chamber holds 50-Site SuperCloner, a propagation system from the manufacturer. Up to 50 seeds or seedlings may be planted on this 50-Site SuperCloner. Two 24 watt T-5 fluorescent bulbs are used to light the cloning chamber. Adjustable shelves are also present which enable easy height adjustment of the plants. Each grow chamber holds up to 16 plants in respective reservoir tanks. Each reservoir holds up to 15 gallons of nutrient solution. That is, the system comes with 'two 16-plant hydroponics system' made of highly durable molded plastic. Air cooled lighting system is used in these grow chambers, using a 400 watt and 600 watt HPS lighting system. The chamber also includes 6" tempered glass adjustable light tubes and reflectors that help keeping the lights always at an ideal height above the plants.

Trinity 3.0 automated hydroponics box system also contains a 'total germination package', 'RO water filter' that produces up to 200 GPD (gallons per day) pure water with low PPM, full lighting system, a nutrient kit, two digital thermometers, a hygrometer, two industrial grade carbon filters, a CO_2 Boost bucket that provides plants with Co2 for a full 90 days, a pH control kit, two air pumps, two water pumps (with a capacity of 185 GPH), two internal circulation fans (4 ½" fan), one TDS meter, two 10 socket industrial power strips, analog double and single timers, two shock busters for eliminating electricity-related malfunctions, a growing media kit containing rockwool cubes and clay stones, one measuring cup, a '24-7 Nutrient Meter', a pair of Grow Goggles, specifically designed to protect human eyes from Ultra Violet and Infra Red rays emitted by HPS lighting system and an Ozonator, to produce ozone which when produced in right quantities destroys many harmful microbes thus creating a clean growing environment.

Price: $4,295.00

5.7 Technology: Automated All-in-One Dual Chamber Grow Cabinet

Brand: Universal Hydro, USA

Product: Yielder Pro XL

Product Overview This grow cabinet contains dual light tight grow chambers – top chamber is used for vegetative growing while bottom chamber is used for propagation purposes. It has a 'double filtered air filtration system'. First filtering is to ensure all air coming in the cabinet is free of bugs, fungus and dirt. Second filtering by a 20 lb carbon odor filter removes odor leaving the cabinet. A fusion of DWC and aeroponics technologies is used for bigger yields. Duct silencer and insulation ensures silent and stealth operation. 100% diffused Mylar coverage ensures light recycling for total coverage and efficiency. Caster wheels are provided for full mobility of the cabinet. A reserve tank and multiple air intake ports are also available with the system, besides all standard components. This grow cabinet can grow up to 14 full grown plants of 3 feet tall, 14 mid-sized plants of 18 inches tall and 72 clones or seedlings.

5.8 Technology: Hydroponics Grow Tent

Brand: Gorilla Grow Tent, USA

Product: Grow Tent

Dimensions – 9′ x 9′ x 6′ 11″

Product Overview: Grow tents are generally used to cover automated hydroponics grow boxes to provide an enclosed or protected growing environment for the hydroponics crops.

Price: $1,195.00

5.9 Technology: Vertical Hydroponics (VerticalPonics)

Brand: SuperCloset, USA

Product: Fully automated VerticalPonics grow system containing Big Buddha Box and Grow Tent

Dimensions – 96"w x 96"d x 84"h

Product Overview: This product is meant for Vertical Hydroponics (VerticalPonics), a vertical growing system that uses hydroponics technology. The advantage of this VerticalPonics system is that a grower can increase the area of crop production by simple adjustments. For example, a 8' x 8' growing area (64 sq. ft.) can be divided into four vertical growing areas of 4' x 6' (24 sq. ft.), thus achieving a total 120 sq. ft. of utilizable growing area. 4 vertical net trellises are provided for plant support.

Central unit of this VerticalPonics system is a Big Buddha Box that can house up to 78 plants. Two air-cooled lights are provided at the center of Buddha Box, enabling the plants to grow towards the light source.

Big Buddha Box

The Buddha Box contains 80w air pump for oxygenation or aeration purposes and a 70 gallon reservoir that enables the Ebb and Flow hydroponics system, key to the success of VerticalPonics. It also includes a 'total germination package', RO water filter, full lighting system, a nutrient starter kit, two digital thermometers, a hygrometer, two industrial grade carbon filters, a CO_2 Boost bucket, a pH control kit, two water pumps, two internal circulation fans, one TDS meter, two 10 socket industrial power strips, analog double and single timers, two shock busters, a growing media kit containing rockwool cubes and clay stones, one measuring cup, a 24-7 Nutrient Meter, a pair of Grow Goggles, and an Ozonator.

Advantages of VerticalPonics (according to the manufacturer)

- VerticalPonics quadruples yield per unit watt and yield per square foot by increasing the growing space available within the hydroponics grow system
- In VerticalPonics, plants grow up to 2 - 5 times faster
- In VerticalPonics, plant height will not be restricted rather plants grow towards their light source, by growing inward at a 45 degree angle (In Buddha Box growing system, 4 vertical net trellises are provided to train the plants and support them as they reach towards the light)
- In VerticalPonics, both quantity and quality are increased

Price: $5,995.00

5.10 Technology: Fully Automated All-in-One Hydroponics Trailers

Brand: SuperCloset, USA

Product: Super Grow Trailer

Product Overview: These are fully customizable trailers and can be set up with fully automated Big Buddha Boxes. The entire unit is 18 Foot in length and includes one Big Buddha Box, and one Ebb n Flow table with an air-cooled lighting system. This trailer is insulated for all weather conditions and crops can be grown throughout the year.

Price: $25,000.00

5.11 Technology: Hydroponics Greenhouses

Brand: American Hydroponics, USA

Product: Starter Package Greenhouses

Dimensions: 22'wide x 28'long x 4' high sides

Greenhouse Features

1. 22'wide x 28'long x 4' high sides

2. Includes framing steel and hardware for each end wall which has 8mm triple wall polycarbonate

3. One 4' x 6' 8" insulated, lockable pre-hung door

4. Double layer of 6 Mil, UV treated polyethylene film plus inflation kit and wire lock system for attaching poly to the greenhouse

5. 2 exhaust fans, 2 intake louvers and thermostats for cooling and ventilation of the greenhouse

6. One 100,000 BTU high efficiency LP (liquid petroleum) or natural gas heater with heater hanger kit, vent pipe assembly and thermostat

Price: $7,488.00

With this starter package, the manufacturer also offers many options for hydroponic systems. Some of these options are,

1. 612 NFT Herb And Lettuce System

    Price: -$1,595.00

2. 210 Vine Crop System

    Price: -$1,495.00

3. 612 NFT Herb And Lettuce System and 210 Vine Crop System

    Price: -$ 3,090.00

4. 2012 NFT Herb and Lettuce System

    Price- $8,695.00

6 Annexure

Annexure 1: Market Price of Some Ready-to-Install (fully automated) Hydroponics Systems

Automated ready-to-install systems	Growing area available (approximate)	Estimated Price in USD	Supplier
Grow Cabinets/Grow Rooms	65 sq. ft.	$5,000	SuperCloset
VerticalPonics Grow Room Trailers	120 sq. ft.	$25,000	SuperCloset
Aeroponics Systems	35 sq. ft.	$1,200	General Hydroponics
Hydroponics Greenhouses/Polyhouses	600 sq. ft.	$7,500	American Hydroponics
Greenhouse Hydroponic Systems -612 NFT Herb And Lettuce System	60 sq. ft.	$1,600	American Hydroponics
Greenhouse Hydroponic Systems - 2012 NFT Herb And Lettuce System	240 sq. ft.	$8,695	American Hydroponics
Greenhouse Hydroponic Systems - 210VC Tomato/Vine System	20 sq. ft.	$1,495	American Hydroponics

Annexure 2: Market Price of Various Components of a Hi-Tech Hydroponics System

Description	Market Price (approximate figures)
Rockwool	Cube (2 x 2 x 1.5 inch) - $3 Slab (36 x 6 x 3 inch) - $8
3-Tray Propagation System	$500
Seedling Heat Mat	$550
Heat Mat Thermostat	$100
Thermometer	$150
Hygrometer	$150
Ventilation Fan	$300
Air Pump/Aerator	$50
Ballasts	$400
400w Metal Halide Bulb	$50
1000w HPS Light	$100
LED Grow Light	$600 - $650
Reflectors	$150
CO_2 Kit	$200 - $250
CO_2 – Filled Bucket	$135 - $150
CO_2 Refill Bucket	$110 - $125
Nutrient Kit	$200 - $300
Ph Control Kit	$5
Ph Meter	$150
EC Meter	$150
24-7 Nutrient Monitor	$220 - $250
Ozonator	$20
Timer	$130
Water Pump	$50
Water Filter	$315 - $350
Water Meter	$150
Fogger	$2,750
Neem Oil	$30
Biofertilisers	$30
Rooting Hormone	$30
Grow Glasses	$100 - $150
30 Gallon reservoir tank	$150

(All prices in USD, as on September 2012)

7 References

SuperCloset. (2012, September 18). Retrieved from http://www.supercloset.com

General Hydroponics. (2012, September 18). Retrieved from http://www.generalhydroponics.com

American Hydroponics. (2012, September 18). Retrieved from http://www.amhydro.com

North American Hydroponics. (2012, September 18). Retrieved from http://www.wearehydro.com/index.htm

Universal Hydro. (2012, September 18). Retrieved from http://www.cabinetgrow.com

Gorilla Grow Tent. (2012, September 18). Retrieved from http://www.gorillagrowtent.com/

www.ingramcontent.com/pod-product-compliance
Lightning Source LLC
Chambersburg PA
CBHW081245180526
45171CB00005B/548